COSMIC EGG

COSMIC EGG

POEMS

BRENDAN McBREEN

MoonPathPress

Copyright ©2017 Brendan McBreen

All rights reserved. No part of this publication may be reproduced distributed or transmitted in any form or by any means whatsoever without written permission from the publisher, except in the case of brief excerpts for critical reviews and articles. All inquiries should be addressed to MoonPath Press.

FIRST EDITION
Printed in the United States of America
ISBN: 978-1-936657-26-1

Cataloging Information: 1. McBreen, Brendan, 1979–; 2. Contemporary American Poetry; 3. Astronomy; 4. Gender Identity.

Book design by Tonya Namura using Gill Sans & Minion Pro.

Front cover art: Cosmic Tulips, 2016, a collage by Brendan McBreen incorporating photography by James Rodgers and images from NASA's Hubble Space Telescope (NASA, ESA, F. Paresce (INAF-IASF, Bologna, Italy), R. O'Connell (University of Virginia, Charlottesville), and the Wide Field Camera 3 Science Oversight Committee).

Back cover art: Cosmic Bird, 2016, a collage by Brendan McBreen using images from NASA's Hubble Space Telescope (NASA, ESA, the Hubble Heritage Team (STScI/AURA), A. Nota (ESA/STScI), and the Westerlund 2 Science Team).

Author photos: Philip H. Red Eagle.

MoonPath Press is dedicated to publishing the best poets of the U.S. Pacific Northwest.

MoonPath Press
PO Box 445
Tillamook, OR 97141

MoonPathPress@gmail.com

http://MoonPathPress.com

We are part of this universe; we are in this universe, but perhaps more important than both of those facts, is that the universe is in us.

—Neil deGrasse Tyson

CONTENTS

I PLUMBING

I feel like an egg	5
snowfall	6
something in the shape of a bird	7
counting backwards from pi	8
let what is invisible sustain you	9
time	10
algebra	11
I should have been a plumber	12
imagine me a stranger	13
I have a very mammaly family	15
convergence	16
space	17
they were testing the intercom	18
Snodgrass	19
There were bats in Parker's belfries	20
Them!	21
blue sky	23
Time and space and gravitation have no separate existence from matter	24
Cento 1	25
castle in the sky	26
naked we run	27
Sasquatch	28
all apples wear their shadows like dark endings	29
even water has a shadow in winter even breath	30
I've never been told	31
Braille sky	32
a moment away	33
a prayer of sorts	35
Voyager	36
wishing on a black hole	39
cosmic plumbing	40

II RAINLIGHT

Fool	43
metamorphosis	44
Hierophant	46
Hanged Man	47
somehow	48
where are you, Nessie?	49
L.U.C.A.	50
Van Allen Belt	52
nor any altars in the sky, 6	53
nor any altars in the sky, 13	54
nor any altars in the sky, 15	56
nor any altars in the sky, 20	57
will the maker of all things step forward	58
in a fix	59
rainlight	60
quantum universe	61
apocalypse with mustard	62
Through the Breaking Darkness	63
Alive	64
first came the flying machine	66
Time Travel: a beginner's guide	67
juicy mantra of sprawl	68
They say	69

III BETWEEN HOLIDAYS

Nine of Cups	73
shadows like wine linger	74
breath	75
shadows linger	79
when gravity fails	80
in high school this guy with a horrible mustache asked me if a friend was having her period	82
radiances	84
Postcard from Cape Meares	86
I think I'll ask Him	87

I don't remember his name	89
blood rockets to oblivion	91
between holidays	93
black (with pink lace roses)	95
you asked about my childhood	98
Again	100
Grandma made a blanket for me	103
the inevitable	105
logic never yells	106
house demolished	107
the serpents	108
my niece was so excited to see a spider	109
I blew a candle out	110
making soup for psychoanalysis	111
Vertigo	113
first snow	114
I left a basket of crow feathers	115
offering	116
plant everything	118

IV SNOWFLAKES

Ace of Comets	121
Page of Forgotten Oaks	122
Queen of Feral Cats	123
Queen of Illuminated Manuscripts	124
a city in June's scar	125
Message From God	126
phantom poem	127
This is NOT a good poem	129
I took my pen apart	131
Sacrifice	132
a skull-shaped goblet	133
Scribe of Sorrows	134
in the waiting room	135
neon pink woman	136
Thief of Whispers	137

candle that burns the brightest	138
Metallurgist of Symbols	139
Shepherd of Needles	140
When life gives you lemons	141
we told the manatees	143
Tri-nocer-angle	144
Muckoguxil	145
Anxiety	147
Flowerer of Folded Linen	148
Grower of Closets	149
Searcher of Lost Light	150
Student of Eden	151
Friar of Magnetism	152
Goatherd of Edges	153
Translator of Moonbeams	154
Ambassador of Flutterings	155
Inventor of Repeating Inventor of	156
Traveler of Neurons	157
Bard of Unknowing Knowing	158
Bard of Pestilence	159
Making Paper Snowflakes	160
Poem Notes	163
Acknowledgments	165
About the Author	167

COSMIC EGG

I
PLUMBING

I feel like an egg
    —Albert Einstein

feeling its way
toward light

a mole
    inventing earth

one little crab
so far
    from the sea

struggling

up a rock mossy shore

I have to let go
    of what I hold on to

    what holds me

it can't be defined

    until it's gone

snowfall

Even age has very beautiful moments.
    —Albert Einstein

space
    is not space

space
    is not empty

space
    is itself
        connected

finite spherical universe

    we are space
    we are time

stick out your tongue

    and taste it

something
in the shape of a bird

it sings

    sky
rearranges itself

beneath trees
light
and shadow
are fluid

the song
is gone

yet

      lingers

**counting backwards
from pi**

I found your smile
folded
between integers

I hope
you were only
marking your place

it would be a shame
to lose
such an important

    expression

let what is invisible sustain you

*Science without religion is lame,
religion without science is blind.*
   —Albert Einstein

matter

interacts

spatially

causing

3-dimensional waves

a laurel branch

    twisting

        reaching

            for the sun

time

time
ticks by
subtraction

accumulation
its own duality
all things move without movement

You may delay, but time will not.
　—Benjamin Franklin

algebra

the light
above my desk
    went out

I should have been a plumber
—Albert Einstein

unclogging
subatomic particles

encrusted

on the elbow joints
of kitchen sinks

in the drain
    dead goldfish
    furious hair anemones
    a quarter
    unknown blue goo

all bouncing
in aquatic Brownian motion

converging
    in emptiness
    in blackness

compressing

where even light
    cannot penetrate

but water knows

and the floor
    is wet

imagine me
a stranger

I consider myself to be a stranger everywhere.
   —Albert Einstein

imagine the wave

center it
   in a spherical lake

standing
   in free space

waves of electrons
   collide

     with paper swans

ripple
   to the outer edge
     of time

how could I ever tell you

I want to be a woman

I want
   to be a swan

rose print paper
petals scattered

some
    on fire

I want to fold myself

       into myself

I have a very mammaly family

except for brother Kyle

who is some type of reptile

and cousin Kim
who has fins

and cousin Gil
who has a snout like a trout

some of us have scales too
some of us are aqua blue
others have covers of feathers
and a claw
like uncle Paul

I guess my family

isn't very mammaly
after all

convergence

The meeting of two personalities is like the contact of two chemical substances: if there is any reaction, both are transformed.
  —C. G. Jung

we
are
all
four dimensional patterns
of chemical reactions

reacting
to one another

bouncing
from mis-conclusion
to missed chance
without a chance
or reason
in this season
this electric year

full of monkeys

throwing politics

at each other

space

space
ether
an expanse

beyond the human
yet including us, a
darkness where we find our light

We are an impossibility in an impossible universe.
   —Ray Bradbury

they were testing the intercom

in the wrong room

I asked, *God, is that you?*

one of them said, *yes*

I told him

if he were really God

he would know
what room I was in

at which point

I was struck by lightning

Snodgrass

was the type of man

who suddenly explodes
into burning bits of ash and sulphur

without even emailing anyone.

He only did it once

but everyone knew

he was that type.

So we avoided him.

It took us two years
to figure out he was dead.

And another six months
to realize we had a volcano problem.

We called the volcano exterminator

and he said it's not unusual,
volcanoes are sneaky,

and some people

are just like that.

There were bats in Parker's belfries

A whole lot of them
echoing back and forth
sending secret messages
in bat-Morris code.

It had to be stopped.
Parker thought they were talking about him.
And the last time Parker thought bats were talking
    about him,
he got married.
We couldn't let him do that again.

He and Serena were happy.
Besides,
Serena is really lucky to have Parker,
she's such a gorgeous little blood-sucker.

Them!

they are
veins
branching
bisecting
out
to reach
    capillary crumbs

or veins
of a leaf
each strand
seeking
    sunlight
      sustenance

these ants

do not seek
    our sun

their light

is deep
    within earth

these ants

construct temples
where they please

these ants

are the body
 and blood

 of their queen

blue sky

Flying is learning how to throw yourself at the ground and miss.
　　—Douglas Adams

blue sky
falls into my arms

we dance like hypochondriacs

finally free
from the pressure
　　of health

Time and space and gravitation have no separate existence from matter
　—Albert Einstein

fireflies

among azaleas

a ripple

dreaming

stars

sparks of dust

in the fog-dewy night

Cento 1

It's the dream we carry
From the Northern lakes with the reeds and rushes,
for the thoughts we share
not ready to give it over
to the eternal surf, to Time!
Swallow this, it will make you well.
Come stand with me, look forward, not back,
For this is love and nothing else is love,

castle in the sky

the eye
watches
the world
spins
faster
until all motion
is one continuous
falling
and rising
hot and cold
combine
spouting
philosophies
of water
and wind
until the conclusion
bounces
safely
away
cushioned
on the same
relentless
air

naked we run

across the sky
gold and silver fishes
holding our breath

Sasquatch

are you
what we would have been

before
the Annunaki

changed us
made us more like them

Bigfoot
North American

great ape
our cousin lemur

are you
our past haunting us

our lost
potential made flesh

made fur
leaving loping tracks

across
our most secret dreams

all apples wear their shadows like dark endings
　—Jessica Goodfellow

all he ever wanted was paradise where
apples ring like wind chimes and
wear yellow spots on
their peel the sun in each core, but
shadows oozed from the other side of the world
like jellyfish or angry phantoms bent on
dark apple smashing ecstasy he was used to
endings but not his own

even water has a shadow in winter
even breath
—Jessica Goodfellow

even the
water
has had better days
a tall
shadow floats
in
winter sunbeams
even a cat can't steal
breath from the dead

I've never been told

to go fly a kite

I have been told
to go play
on the freeway

I'd prefer
a kite

like fishing

for god

try to reel
him
in

show him off
stuffed
lacquered on the wall

I once caught a god
(this big!   )

but he got away

the line broke

while I was dodging
traffic

Braille sky

fingerprints

of blind gods

a moment away

time
mystery
what is
but is not
understood

what is understood
but still amazes

what amazes
the eye
delights the dark

the time
will come
when time
becomes

not understood

not amazement

not mystery

but mystery
beyond mystery

lightning
frozen
in a moment

purple glows the clouds

I can see
static

the falling rise

of Earth

reaching heavenward
for mystery

yet time

is gone

devoured

amazed that
it can be devoured

yet dark
light
purple
and strike

plant seeds still

frozen
in the mystery
    of ink

a prayer of sorts

whoever we are
to become who we might be

deliver us
from safety
free us
from fear
and conformity

bring us peace
in the chaos
nature provides

let control wither
order classify itself
to extinction

praise
every apocalypse
as rebirth
second chance

sin expunged
forgotten
in the rhythm and flow
of living

Voyager

long after
our wars
and our pollutions
fail to kill us

humans
and the Earth we cling to

will return
to the dust
stars are made of

no whispers
no bangs
just cold heat
from a dying sun
a final embrace
of light

the small hopes
we have
scattered
into endless oceans
of time

nothing more
than graffiti
on the universe
saying only
we were here
we existed

even these
will be lost
among gathered dust
and debris

until they burn away
streaking
across the sky
of some distant planet
born eons after
Earth's demise

a creature there
may find
an odd relic
and say to its fellows
these metals
these etchings
cannot be made
in nature

amid phosphorescent
flickers of dissent

this creature

will be laughed out of the room
into cool night air
where it will tilt
its sensory protuberance up
to better hear

the stars
sing

and think
maybe
just maybe
we are not
    alone

wishing on a black hole

before the ender of time

finds us

wishing
    we had more

before all dust

is returned

to its singular source

let us add

a few splashes
    of color

to the unfinishable canvas

of understanding

cosmic plumbing

swirling
along the event horizon
of a black hole

time
is how
we perceive
    spaghettification

our past

constant draining

anchoring
our future

we run
as fast as we can
just
to stay
    put

the final flush
imminent

and what manner

    of celestial alligator

might we meet

on the other side?

II
RAINLIGHT

Fool

no rules

hold

this is just
the beginning
middle
end

it's the same

this journey

only one step

beyond

sky sleeps

coyote calls

the moon

is hidden

    from view

metamorphosis

do you remember

you who devoured yourself

yellow stripes
flicker
in azalea time

drinking
sucking
nectar

the life of flowers
given
for future's sake

for young
to be born
to blossom
to pollinate
other butterflies

but you
oh monarch of meanderment

do you recall

the days
and weeks

digesting
that noble worm

unknowing
the price of wings?

Hierophant

words

make themselves

known

to wings

to wind

now

there are only letters

now

incoherent marks

soon

they fade

all that's left

the listening

Hanged Man

try trading
comfort
for strife

try giving blood
to a stone

blink
and think of balance

would dropping
lift you higher?

would falling
make you fly?

somehow

we forgot to breathe

just held our breath

until we floated away

the world below
looked
like such a good idea

so we fell

and kept falling

not letting the ground
    stop us

where are you, Nessie?

submerged
in endless doubt

traveling
the Paleolithic
    and back

between
camera clicks
just missed?

are you
in cardboard model heaven?

forward looking sonared
to distraction

tired
of the limelight

tired
of doubters and fakers
the wannabes and has-beens

did you tunnel
to the Atlantic?

catch a Sargasso ferry

to Mars

    and beyond?

L.U.C.A.

the first
    is Last
the miniscule
    is Universal
the extinct
    is Common
the foreign
    is Ancestor

divided
among multitudes
that devour themselves
    defend themselves
you were legion
    primal
in a dark ocean
across the ooze
    of time
beyond the touch
    of sight

floating

    devising

a chromosome for this
stem cell for that
    chloroplasts
    mitochondria

Gestalt unity
of purpose

harmony
    from bedlam

you were a single cell god

yet demanded nothing

forgive us
we have forgotten you

our evolution
    biology
        genealogy
our heritage

our truth

has replaced
    your own

we began
    with chaos

and to chaos
    we have returned

only the water

in our ribosomes

    remembers you

Van Allen Belt

planets sing
to each other

Earth
sings
Humpback whale song

water music

the sound
of fire wind
redirected
around
water magnet

is it any wonder
for humans

who think ourselves planets

that music

is in our blood?

nor any altars in the sky
 —Lawrence Ferlinghetti

6

oh great and powerful Oz!
    bless this bounty we are about to receive

give us strength!

    to smite
    the neighbors
    whom we love

forgive us our sins

and burn those nonbelievers
    in hell
        roast them alive!
    with searing maggots
        eating eyes
            and entrails

until they learn
    that YOU
        are the one TRUE god

god

    of love

        and of peace

nor any altars in the sky
—Lawrence Ferlinghetti

13

belief in punishment is punishment,
belief in sin is sin
—Ursula Le Guin

the only god

is the God

    Imagination

we are all her acolytes
in this
    bone and cellular
temple

in air
    we fly free

because She IS flight

we are free
from the hell and hierarchy
doubters
    call faith

free
    and swimming
through mitochondrial imagination

science
    is beauty

biology
    is beauty

brain cells   are beautiful

I want to sing for you
    a molecular love song

where wolves chase
    through misty woods

and the ocean

    and the waves

        and the sky

            and the stars

                and the mind

are all
    there
        is

nor any altars in the sky
  —Lawrence Ferlinghetti

15

we
mosaic
minded
mammals

call it god

but there is no god
    in this machine

only flesh
    and waves
of tangled space
    time

we are
    acrobats

waiting

for the empty air
of existence
    to let loose

so we may flail
    and fall

nor any altars in the sky
　　—Lawrence Ferlinghetti

20

I fell in love

　　with unreality

when reality

became too unreal

　　to bear

because
not yelling
　　frightens me more

because
　　being haunted

is never

a choice

will the maker of all things step forward

there's someone here who would like to talk to you

I've a small problem
    with some of the defects
        and defectors

associated with this life

clovers in a field
    daisies
a pile of rocks to climb

Benjamin Franklin said
little strokes fell great oaks

why would anyone want to harm an oak?

acorns cry
    lament a father's passing

would the maker of all things please step forward

I'd like to exchange this winter

    for a cat

preferably one with short fur

    easy to carry

in a fix

light through Venetian blinds

throws frantic I Chings

on the walls of God's workroom

let them become oil…
so the next semi-sentient species
has something to kill each other over…

but He can't take His omnipresent eyes off
    the soon to be molten sphere

so He tries again
but still the planet is beyond the reach
    of God's mighty hat

He pulls out all His rabbits
    and colorful linked scarves in vain
        hoping for a miracle

for some sort
    of human intervention

He prays
    sighs
    curses Himself

for not being all-powerful enough

    to do it

        alone

rainlight

in music

it's the empty space
    that matters

in Japanese flower arranging

Zen painting

the curve of hips
    wet with rain

ripple
    of bare feet

dancing

    pausing

        dancing

the hollow

    between

quantum universe

the birds

the pond

ripple

apocalypse with mustard

the man in the giant hot dog costume
is the only sane person left

    and nobody wants his coupons

Through the Breaking Darkness

through time
    beyond human memory
the ocean
    breaks sand to atoms
foreshadows
    darkness

we have seen our sins
    light the night sky
        like nebula
we can still see
    our original goal

        to be God

over desolation
    to absolve chaos
implement our own brand of order

then watch

    impotent

        as the ocean

washes

    it all

        away

Alive

millions of tiny
water molecules
converge
with the urge
    to make the world
whole again

was it God
    or were there many gods
all floating
as if gravity itself
    were turned off

should we believe
    all to be created
was once water
    frozen to itself
showing us
    the way
        through time

it would be a pity
if God
    or The Gods themselves
didn't know
    what it was all about

can our future
    be ice
        silent
    motionless

or just cubes
    dropped

    into
    our longing
        cups

first came the flying machine

then

snow   snow
    snow   snow
       snow

eighteen feet of it

houses disappeared
into buried ruins
no archeologist
    could ever find

without internet
people panicked
and formed
whole new
mole tunnel societies

children
who grew up there
drew pictures
    after the thaw
of helicopters
    with icicle teeth

Time Travel: a beginner's guide

late
late
late
rushing here
    there
into this room
    back to that one
out the door
and back
    to the bathroom
just keep telling yourself
    you're wiser
out to the car
and back
    for your glasses
back to the car
    out through traffic
late late late

I'm sorry sir
your appointment
    is next week

juicy mantra of sprawl

the old ones spoke of this
    they said:

        run

They say

you should reach for the stars

        that's nice

but under city skies
    there are flashing lights—
        police cars
        ambulances
        fire trucks
    unhelpful helpful lights—
        stop lights
        street lights

with all that ambient light
    you can't always see the stars

all you can do
    is reach for the darkness

        and sometimes

            it's all you get

III
BETWEEN HOLIDAYS

Nine of Cups

wealth
    seeks
    itself

the mirror
is only
a reflection

to be

one must

answer

even
when the question

is forgotten

shadows like wine linger

on the edge of dreams
I long for reason
to try again
ten thousand times
I sink before I swim

breath

fresh
and sweet

but the walls

are falling

and I am

but I don't know why

where are you
who was me
where
is time

in my hands
I flex
all I have known
drink
dreams
and begin

begin

begin anything

any something

so long as it is new

nearness fades

pretend

we are dancing
there are flowers
beneath our bare feet

and the walls
never fell

am I looking in a mirror

the mirror

there's only one

it's kept
behind glass
in a forest
where sea creatures
pretend
they are lungs

I breathe

and softly
slowly
ground
rises

meets

the walls

I want to see beyond
through them

see myself
looking back

looking

beyond me

but my eyes
are the eyes
of blood
and carbon

they burn phosphorescent
when in need

they see brick and stone
as
brick and stone
with dragons dancing
between

but only brick and stone
and dragons

when we met
were you
what I thought
I would be

the walls
have been here

longer

than me

or any other me
before sky and water
they were a forest
and starlight

out of mist
they became
something
not mist

I can feel you

reaching

there are no
windows
in these walls

but I can't
touch them

when the walls
fall

where will I be?

shadows linger

the wind
is gone

gulls cry

a knife rusts

a heart

can only take

so much

when gravity fails

Spotsy
my grandmother's cat

adopted me

Tiger
from her latest litter
was my best friend

we chased each other
all around our back yard world

he hid in the wood pile
liked the smell

I'd reach in
he'd bat my hand
as I tried to pull away
he'd nudge out
I'd bat at his paws

then he was gone

my dad said
he asked my brother
to drive him to the country

and leave him there

he found his way back

with a broken hip

they took him somewhere again

he didn't come back

in high school this guy with a horrible mustache asked me if a friend was having her period

what the fuck business is it of yours, I thought
I don't know, I said
"Cause you know she didn't talk to me in math this morning
so you know I thought that might be it—"
she doesn't like you, have you considered that? I thought

I heard her talk about being in his house with some friends
how he drank huge gulps of milk from the carton
belched
wiped his mouth with his hand
later he grabbed her hair and tried to kiss her
she pushed him away

he mentioned all this in gym
said she was all over him
she kissed him
it's only a matter of time before she goes all the way

I saw her in history
something was wrong

in the hallway
I heard her
talking with another girl

her little sister was in the hospital

a car accident

they hugged

both
crying

radiances

may it come that all the radiances
will be known as our own radiance
    —Tibetan Book of the Dead

there is beauty

    in death

reaching
    of bleached skull antlers

pine cones
    brown leaves
    rusted chain

long unseen
    cobweb

aren't we all
only half alive anyway?

or less?

    dead hair
    dead fingernails

most cells
    already
        gone

outside

   a crow
      watches

Postcard from Cape Meares

the waves are foaming themselves on a shore I can't see

gulls and other sea birds silhouette white gray wisps
and scratches

diving and swooping across brush stroke nimbus building
cumulus and occasional blue sky

I am listening to the steady twitch of a clock on yesterday's
time while the sun is undecided about whether or not to set

moments may fade faster than we can realize they exist

but right now in this already gone moment there is
something which will not fade

something of the sea

never the same sea from one moment to the next
but always it's still the sea

I think I'll ask Him

the tattoo
along your spine

tells the story

of our creation

we were made
from a spare lung

left over
from Adam

from Lilith's tears
at his betrayal

it all worked out though

Eve
followed a trail
of apples
left by Lilith

they fell madly in love

moved to Bethlehem

they write poems
sell pottery

Adam

he has only himself

to play with

I don't remember his name

i

I knew a kid early in high school
he tried to be a bully
wasn't very good at it

he threw bits of paper at me
the teacher told us to knock it off
I pointed to spit balls around my feet
none near him
he called me a traitor for that
I could never convince him
I wasn't on his side to begin with
he called me a faggot

behind his back
people talked about him
said he liked to start fires
hurt small animals

he could sing
soprano
he sang at an assembly
we tried to make fun of him
but couldn't

he was too good

ii

last week
I saw his picture in the paper
he tried to burn a church
only scorched one pew
then he cut his balls off
stabbed himself in the heart
    missed

I wonder
    can he still sing?

blood rockets
to oblivion

when the first one sang
the others
    flew away

*

what do you plan on doing
    with your life?

*

I don't need sleep
a few hours
    some silence
    stars

*

since I was conceived
everything has gone wrong

*

someone else's reflection

I'm stuck being male
trying to make the most
    of a bad situation

*

life is just one deep enough cut away

*

they say astronomers live in the past
and time
is the unintended consequence
　　of falling

*

there are no highs
but some lows
　　are lower than others

*

I want to feel the warmth
　　inside you

between holidays

with dad
in one
dusty blackout thick edge despair knife apartment
or another
    with a pool table

air
always
pungent

dad watches TV
eats TV dinners
TV desserts

licks static from the screen

dad didn't use his tongue for talking
TV did that

he complained though

bitched and swore

about bills
paperwork
not enough time

    I never suggested
if he stopped watching five hours of TV
when he got home

he might have time for other stuff

    I never said that

I stayed in my room

    headphones
    turned
        way up

black (with pink lace roses)

I've been trying to kill myself since I was twelve
this time
I'm sure to succeed

we're moving
while packing I find an old box cutter
never used

moonless autumn night
careful and quiet I arrange
    a shrine
on the desk by my bed
a ballerina figurine found at a garage sale
for a quarter and a funny look
the postcard with Spanish Blue Bells
and a powder blue butterfly
fragments of shells worn smooth
mom and I found on the beach

then the note
always placed on top the words
I should be female
etched long ago into caramel brown wood

I never write suicide notes
instead I leave obscure poetry:

> *she of winter wind*
> *when her breeze touches your soul*
> *you will understand*

everyone asleep
it's time

I lie down naked
a folded white towel under my left wrist
the box cutter pressed into the green-blue line beneath
skin
I clench my teeth
pull the blade across

    not deep
enough

I stifle fluttering in my chest
try again

again

again
one layer
    at a time

amazing
how little blood
for so much pain

five a.m. I give up
curl
into a ball
pull out fistfuls
    of hair

there's no way I'm going to school today
or the next

so Saturday
    moving day
I have detention

my father grumbles
I'm at school all morning
when I should be helping move

in study hall I draw mushrooms fairies and feathers

back home
my brother and his grinning friend are outside
    with my desk
they look at me
laugh

the rest of the day carrying boxes
they stare
    whisper
        laugh
I avoid them
    the best I can

I'm afraid
if they look
in my eyes
they'll see

the underwear

    I am wearing

you asked about my childhood

divorce
was a cyclone
my mother
my anchor

dad came home (not his home anymore)

I didn't want to go

but wind
spiteful indignant fangs
tore through all the defenses
an eight year old could muster
he's mine
cursing cords hissed

held by waist over his working shoulder
I clawed pictures off the wall
something to hold on to
my anchor
powerless

tossed into the truck
I cried
I couldn't cry

we stopped
to gawk at an accident
dad said it would cheer me up

a silver car at the bottom of a hill
upside-down

people were watching
I think they were all smiling
I couldn't imagine why

there was a blue policeman
telling people not to get too close
the slope is slippery
the ocean deep

I wanted to go home

the typhoon my father was
swept me away

I don't know

if I'll ever
    land

Again

heat
or
another
naked amen

floating
in the distance

    pretending
to be water

**

Pretending to be water
I take
my vows
away
from those who believe
in the sanctity
    of intolerance

I call them
by their real names

no one hears me
    they think
I am mad

**

They think I am mad

of course I am mad

how many double standards
does it take
to make
the right

    right?

**

We hold these truths to be self-evident…

    some selves
    count more
    than others

That all men…

except if they are Black
    or women
    or Indian

except men who love other men

except

    except

        accept

**

I will not
accept

    this

**

again

we find ourselves

a nation
afraid

of the change

that
has already
taken
    place

Grandma made a blanket for me

the color of fall leaves

I wore it like a cape
and ran around the house
pretending to be a who-who bird

I wasn't supposed
to play in the living room
so that's where I went

it was quiet there
I liked the big window
with pink rhododendrons
the crack in the glass
that drew rainbows on the floor

I'd run in circles
waving my wings
shouting
    who!
    who!

the cat
found this behavior amusing

I'd land
crouch down
cover myself
with blanket wings

nothing could hurt a who-who bird

later
I'd curl up in the sunbeam
next to Fluffy
a Balinese
my grandmother's favorite
and whisper
    the big secret

Fluffy yawned
    and went back to sleep

the inevitable

has a way of happening

like when you're falling

the ground
getting bigger and bigger

then

    the tide comes in

logic never yells

reason never shouts

    but you say

        you're not yelling
        you're not shouting

you're screaming

    and that makes it okay

house demolished

years ago
nobody told

    the tulips

the serpents

in my eyes

have not

separated

my eyes

from

all eyes

I have seen

my niece was so excited to see a spider

look Mama a spider!

then Mama squashed it

she pouted for two days

now

she's helping me clean

and there are two spiders

I point to one
    and ask

should we put it outside

she squashes it
    twists it deep into the carpet

I don't tell her
    about the other one

I blew a candle out

and left the room

when I came back

there was a mosquito
    drawn

by my breath
    to molten wax

frozen

    in its first

    drink

making soup for psychoanalysis

add self-doubt
    for realism

include the bone
of a recently waxed
    octopus

or
if octopi don't have bones

use a screwdriver

preferably Phillips

you will need the iron

and something solid
to complain about

add a five leaf clover
    not for luck
but because it's green
and green
is your new favorite color

if only for today

add the wish
of an autistic child

it needs a little sweetness

now add every bitter argument

you have ever had
with your father

let it boil for ten minutes

before deciding
    on carry-out

Vertigo

living inside a cheese grater
smashed against
other people's expectations
the bridge from point 'A'
to point
wherever the hell
my destination ends up being

first snow

houses
without children

I left a basket
of crow feathers

on your porch this morning

dawn shined through me
filled my hollow soul

    with footprints

I had forgotten

    how to hold you

offering

I offer
    my pound of flesh
wherever I go

there have been
    no takers
        so far

in the season
    when rain
is the way of things

the sky
is not always gray
    but sometimes
        full of seagulls

wings spread wide
completely still
    as wind
transforms gray theologies
    around them

cries linger
few by few
    they fall
away

until the last one
becomes
    the north star

this is where I orbit

offering
    broken bits of myself

to any soggy seagull
    I meet

plant everything

that hurts

the earth

will forgive you

IV
SNOWFLAKES

Ace of Comets

leave behind
ice
for memories

reflect
more than

yourself

create
a life

between

fallen

    trees

Page of Forgotten Oaks

simply
trying
is sometimes trying

in moonlight
all silver
is water

and the wind

full of salmon

never stops
swimming

uphill

Queen of Feral Cats

a pushy lot

always
wanting
in
and out

or

just curious

about why

you're standing there

with the door open

Queen of Illuminated Manuscripts

lines
intersect
swirl
    to life

a griffin

trapped
in a tapestry
of vine

living on paper

one must be wary

of erasers

a city in June's scar

aware of awareness
sifting between now and then

a collection of moments
says to me
shadows disappear without light

torrents glare and pulse
in the great scheme of memes

I think I am dreaming
because every time I try to write
I am awake

but when I wake
nothing is written

Message From God

I received a message from God

in the alphabet soup and vodka I threw up yesterday

it spelled

Zthufalupik

I don't know what that means

phantom poem

on the way home
I had a poem amputated

I don't remember much

being in the truck
waking to surgeon's masks
scalpel slicing paper
    my pen on the floor

nurses told me it was gone
there was nothing they could do

I still feel it sometimes

there was something about
me being autumn wind
    or maybe snow
and the first time I cried in public
    I was unashamed

      no one noticed

the rest is hazy

is my poem still there
among hazardous needles
and severed participles?

    was it incinerated
deep within the bowels of the hospital
a dark furnace with ember eyes
    that devour little plastic bags
    marked with red warnings?

has my poem
itself become the color of autumn?
    with leaf embers
    falling up
trying
    to rejoin their trees

This is NOT a good poem

this is a bad poem

this poem
    is mean to puppies

this poem catches flies
puts them in spider's webs

this poem trips the elderly
glues quarters to the floor
    laughs
when people try to pick them up

this poem is unruly
it heckles other poems
this poem is not a good role model
this poem drinks
smokes
hangs out in tattoo parlors
on the mean side of the tracks

this poem scolds kindergartners
laughs at others' misfortune
this
    is a bad poem

once
in grade school
this poem beat up that slow kid
    and took his lunch money

things have only gone downhill since

this poem is packing

it's a rough neighborhood
this is a rough poem

sometimes
this poem deals pot
rolls junkies and drunks

a good poem wouldn't do such things

this poem is a loner
no need for influences
    Keats
    Nash
    Kipling
    McClure
this poem is self-sufficient
    this poem is proud

someday
this poem will be gunned down on the street
    by grammar police or worse
    but today
no one
messes with this poem
this
    is a *bad* poem!

I took my pen apart

to give it a fresh start
the spring
is a thing I have never seen
that resembles so much a heart

Sacrifice

I offer this sacrifice to you o muses
let this ink
be your blood
let this life
be siphoned
until you
are replenished
let the dry husk
be discarded

they like my pens to suffer
to bleed
    slowly
        dry

I must be careful though

if one pen escapes
before
it is finished

the muses

will be angry

they will make me

write
    embarrassing things

a skull-shaped goblet

catch words with butterfly nets
make tea for my muse

Scribe of Sorrows

an abundance
of confusion

pins
needles
locusts

a cloud of
silver dragonflies

wiffing by

but who needs insects

when the web

is only

half finished

in the waiting room

of the endoscopy center
a book of mazes

neon pink woman

clothes and hair, so mad at the
bee following her

Thief of Whispers

wishes
never grow old
though
wishers do

a sunlit meadow
full of spring
flowers and bees

will always
be home
to more
than can be seen
heard
or
predicted

some wishes

are simply waiting

to be found

candle that burns the brightest

gets reincarnated
as a lighthouse

Metallurgist of Symbols

    slow

deliberate

    move

    as

a cypress

    grows

then chant
    ten times

all
    the names

you wished for

    as a child

Shepherd of Needles

ice
tumbles
into plastic
pings
tings
and plunks

I've heard
being normal

is over-rated

When life gives you lemons

the grass is greener

keeping the doctor away

the mice will

make mountains

out of the frying pan

into a gift horse's mouth

but don't cry over

all the tea in China

because the bigger they are

in glass houses

the harder they fly a kite

without a paddle

until the cows come home

in sheep's clothing

remember

when the going gets tough

an early bird in the hand

dies by the sword

measure twice

and when in doubt

panic

we told the manatees

we were mad
at all the evils
    in the world

they told us
they were angry
at all the Elvises
    in the world

it was clear
we were talking
    at cross porpoises

Tri-nocer-angle

rhino-sosceles triangle
tri-sosceros rhi-angle
you know
one of those triangular
rhi-angular
tri-soscoles
rhino-hedron
tri-noceros
rhina-trina-isoscele-ceros shapes
with the horn and such

or

how I flunked geometry

Muckoguxil

Indications:
    For temporary relief of acute hypochondria

Drug interaction warning:
    Don't take Muckoguxil if you are currently or have ever breathed oxygen

Side effects are generally mild and may include:
    Mood swings abnormal hair growth blurred vision blindness loss of hearing repetitiveness repetitiveness drastic weight gain spontaneous total weight loss hiccups heart problems a third arm may sprout from your forehead death slow death slow painful death fin rot hallucinations about eggs homicidal and cannibalistic urges facial twitches the earth exploding uncontrollable loud diarrhea sexual side effects cancers monkey sounds when you blink invisibility purple urine swelling of the feet sweating dry mouth headache hives melting skin paranoia dyslexia people following you excessive drooling vampirism insanity baldness turning into a frog turning into a six-eyed eight-legged frog fungal growth demonic possession yellowness around your ankles itchy anal rash delusions of grandeur depression giddiness feather crotch and projectile vomiting
    If side effects persist for more than ten minutes contact God

    Muckoguxil is not for people under the age of one hundred and fifty

Do not take Muckoguxil if you are a carbon based life form or are planning on becoming one soon

Muckoguxil
    Live your life!

Anxiety

Do you feel anxious?

Are you generally nervous or tense?

Don't you wish you could face living without being perpetually scared senseless?

Is your neighbor hiding weapons of mass destruction in his rhododendrons?

Try America's number one non-prescription treatment for anxiety disorders:

Buy a gun.

Studies have shown that owning a gun can significantly reduce stress and help calm your fears about the many horrible things your neighbor is plotting to do to you.

Owning a gun is not a replacement for prescription anti-anxiety medications, but it can make procuring your medications without payment easier.

Owning a gun will not help people whose neighbors own bigger guns, however buying several guns of varying sizes may still be beneficial.

Owning a gun is not recommended for children under the age of three unless they pay cash.

Side effects are generally mild and can include bullet wounds, lawsuits, property damage, and dry mouth.

Flowerer of Folded Linen

follow the lion
pass
a ship

fold

yourself

bend

between blossoms

find
the only friendly
    face

in the mirror

borrow
every broom
offered

bow
and bid
them all
good yesterday

Grower of Closets

give
yourself

a melody

sing it
while sleeping

memorize
every maze
inside your head

follow
the path
of least
residue

only use mathematics
to calculate
    never

Searcher of Lost Light

only darkness
is without

wanting

forever
flows

slower

in moonlight

even shade trees
bend
    to music

offered
    by fire

Student of Eden

prognosticate
tomorrow

will the sun
rise

set

burn
ten million times
    brighter

if you are listening

tell me

where is tomorrow
forever
the next moment

below

all is falling

except
    me

Friar of Magnetism

seven
times seven
times
seven

is not the relevant
number

try again

this time
with tulips
pink and yellow

if that doesn't work

begin weeping
until
dark

begin

and end

in the middle

exist

Goatherd of Edges

climb
down

the meadow

is in your mind

follow
passion

lost in undergrowth

when mathematics
is forced
upon altruism

the result

is perplexing

Translator of Moonbeams

forest
    below

glows
and goes
where
forests go

by lamplight

a spider

builds

itself

    a prison

Ambassador of Flutterings

join me

in the dream

    of corn

ankle bells

quetzal feathers

join me

in an Aztec dancing
    dream

be my sun god

remove

    my heart

Inventor of Repeating Inventor of

suddenly
there was the word

suddenly

fragments
of sound
clangled
down
into a profusion
of
suddenly

then the spiraling
all of a sudden
stopped

all of a sudden

Traveler of Neurons

quick!
become more!

not that more
is always better

sometimes
less
is more
important

flying through water
penguins
become
lighter
than the sum
of their
surroundings

fish
don't fish so well

Bard of Unknowing Knowing

in between
flowering moments

bees pause
look
to the sun

dream
that some day

the buzzing

will stop

Bard of Pestilence

information
about
finding
one's self

is rarely
printed
in phonebooks

but when it is

be sure
to keep a copy

especially
in moonlight
when it's cold
and you need

kindling

Making Paper Snowflakes

1. Fold the paper in half diagonally to make a triangle.

try folding night
make a mask to hide behind

2. Now, fold the triangle in half so the pointy corners meet.

the pointy corners
should never meet
but they always do

3. Fold your triangle in thirds...

and keep folding
keep mangling yourself to fit a dark little box
no stars tonight

4. Cut across the bottom of your paper so it is straight.

and keep cutting
cut deeper
cut cleaner
remove any excess
any feelings you may still have
any hope

5. Cut your folded paper so it looks like the triangle above.

like every other triangle above

6. Then unfold it gently.

it's all you have left

POEM NOTES

"Cento 1": A cento is a form where each line is taken from another poem.

    Olav H. Hauge, *Borealis*, translated by Robert Hadin

    Andrew Barton Paterson, *The Black Swans*

    Nikki Giovanni, *A Poem of Friendship*

    Naomi Shihab Nye, *Boy and Egg*

    Pablo Neruda, *Always*, translated by Brian Cole

    Ogden Nash, *Adventures of Isabel*

    Oodgeroo Noonuccal, *Let Us Not Be Bitter*

    Robert Frost, *A Prayer in Spring*

ACKNOWLEDGMENTS

Thank you to the following journals and anthologies in which previous versions of my poems have appeared.

The American Drivel Review: "Message From God"

Aphelion: "Sasquatch"

Avocet: A Journal of Nature Poems: "I feel like an egg"

Bear Creek Haiku: "wishing on a black hole"

Black Petals: "first came the flying machine"

Circle Show: "I've never been told" and "Van Allen Belt"

Crab Creek Review: "house demolished"

The Delinquent: "imagine me a stranger" and "counting backwards from pi"

Espial: "they were testing the intercom"

Eternal Haunted Summer: "Sacrifice"

The Far Field: "imagine me a stranger" (second publication)

Farther Stars than These: "Voyager"

Five Willows Poetry: "Cento 1"

Flutter: "radiances"

Four and Twenty: "plant everything that hurts"

HA!: "phantom poem"

In Tahoma's Shadow: "will the maker of all things step forward"

Kind of a Hurricane Press, *Emergence:* "L.U.C.A."

Kind of a Hurricane Press, *High Coupe:* "a prayer of sorts"

Kind of a Hurricane Press, *Mind(less) Muse:* "when life gives you lemons"

Kind of a Hurricane Press, *Shattered:* "you asked about my childhood"

Kind of a Hurricane Press, *Tranquility:* "something in the shape of a bird"

Mad Swirl: "Metallurgist of Symbols"

Mirror Dance: "Hanged Man"

Mithila Review: "where are you, Nessie?"

Origami Condom: "Muckoguxil" and "my niece was so excited to see a spider"

Raven Chronicles: "Anxiety"

I am grateful to the Helen Riaboff Whiteley Center at the University of Washington's Friday Harbor Laboratories for a residency in 2009 where I wrote a series of poems based on Albert Einstein quotes, many of which appear in this collection.

I also express my gratitude to Lana Hechtman Ayers for her editing expertise and friendship, to Jim Teeters and the Striped Water Poets for years of encouragement and good ideas, Paul E. Nelson, my first poetry mentor, who still challenges and inspires me, Christopher J. Jarmick for his enthusiasm, his vast collection of writing prompts, and his humor, David D. Horowitz for his dedication and kindness, Marjorie Rommel for her insightfulness, Connie Walle and the Puget Sound Poetry Connection for their support over the years, and to many others too numerous to note.

ABOUT THE AUTHOR

Brendan McBreen is a poet and workshop facilitator with Striped Water Poets in Auburn Washington.

He graduated from Auburn Riverside High School, attended college at Green River Community College, the University of Phoenix, and Monash University in Australia. Once long ago he earned his black belt in karate from the American Karate Escrima Association. As a regular attendee of various poetry and literary events, (such as Burning Word, the Skagit River Poetry Festival, the SPLAB! visiting poets series, and Poets in the Park) he has taken classes presented by many notable writers including: Wanda Coleman, Diane di Prima, Christopher Howell, the incomparable Jack McCarthy, former Washington State Poet Laureate Kathleen Flenniken, Elizabeth Austen, and many others.

Brendan is an event coordinator with Auburn AugustFest and the Poetry at the Rainbow Café reading and open-mic series. He is a former coordinator of the August Poetry Postcard Project. And his cooking has been known to make people cry.

He writes sci-fi poetry, haiku, is a surrealist, a humorist, a Seahawks fan, an admirer of crows and cats, and a life-long collector of weirdness. He is a student of Zen and Taoist philosophy, Wicca, psychology, and various other esoteric ideas. He is a collage artist, an occasional cartoonist and sometimes still paints.

Brendan has been featured at various regional venues and published on T-shirts, as graffiti in restroom stalls, in the UK, and in numerous journals and anthologies.

For more information visit:
www.sites.google.com/site/terralunapoetry